The Future of Electrical Generation

The Future of Electrical Generation

This book is dedicated to the family of this earth and their continued existence

The Future of Electrical Generation

Index

Page

4 -------------------------------------- Introduction

9 --- Compressed Air: its production and power

16 ------ Hydro Electrical Generation using Dams

20 ------------------------------ Rain water mounds

27 ------------- Taylor ram pump and its adaption

41 --Check valves for Water and Compressed Air

48 ---------------------- Volumes of water involved

51 ---------------------- Tesla turbine and generator

54 -------------------- Production of distilled water

55 ---------- Siphon system for transporting water

58 ------------------ Air electric system in operation

62 ------ Advantages of Air Electric System over a

Hydroelectric Dam

The Future of Electrical Generation

Introduction

Climate change, global warming, world catastrophes -

The headlines of many news articles and internet sites with their predictions of doom and gloom seem to give mankind a sense of hopelessness and that our planet is coming to an end, yet if we are responsible for any of the above scenarios then our ingenuity should be able to find solutions.

If the use of fossil fuels is creating so much carbon dioxide and other chemicals in our atmosphere that they are changing the seasons and climate, wreaking havoc from earthquakes, hurricanes, tsunamis, tornadoes, rising sea levels and the like, then it is time for a change in thinking. It has long been known that electricity in and of itself is a completely non-polluting entity, but it is its methods of generation that are called into question.

When we use coal, the main stay of the industrial revolution, for the generation of electricity we are indeed causing many health problems as well as creating large volumes of undesirable gases. Even with the introduction of "clean coal", if it could be

The Future of Electrical Generation

cost effective, it is only a band aid solution to the problems of using coal.

Natural gas, in and of itself, is a good solution to helping the problem of harmful gas emissions but with the mining of such gas, using the present practice of fracking, demanding large volumes of water mixed with chemicals, the advantages of such production are minimized by the mining of such gas.

Oil, the driving force of our vehicles and many industrial complexes, has been proven to have many problems when it comes to environmental concerns. Its use as the source of our plastic age has also produced a waste problem that has global implications.

Nuclear power, once thought to be the solution to all our problems, has also proven to have dire consequences when, through either natural or manmade disasters, creates a threat to the health of mankind through radiation.

Yet there are many ideas that have been labelled "renewable", such as ones using sunlight, wind, rain, tides, waves and geothermal heat, have created many new markets around the world. Many countries are putting large sums of money into the science of fusion to duplicate the power of the sun.

The Future of Electrical Generation

Yet for all the present solutions, there is always the reluctance to change, the political balance of power and the limitations of such renewable energy due to geographical locations.

One solution that seems to be generally overlooked is a resource that is virtually non-exhaustible, totally environmentally friendly, and available anywhere on our planet; air. Although it is being used as wind power, its most useful properties are when it is compressed.

From the earliest form of the bellows, mankind has used air as a resource to forge iron, run clocks, and more recently to drive pneumatic tools, such as jack hammers and many other useful items. In the generation of electricity there are plants in both the countries of Germany, and the U.S.A. which use compressed air mixed with natural gas to run large turbines. The idea of storing compressed air in salt caverns, known as C.A.E.S. (Compressed Air Engineering Systems), has been explored and used. Such air is produced by electrical generators when the demand for grid electricity is low, and thus cheaper, and pumped into large salt caverns. When the demand for electricity is increased this air is released and then runs the same electrical generators that produced the compressed air.

The Future of Electrical Generation

Although this is innovative and a good use of cheaper electrical power it is not a sustainable solution to the world's electrical demand.

A similar system has been employed by pumping water into a lake at a higher elevation and releasing the kinetic power when the higher demand for electricity is required, thus using cheap electricity to pump the water into the reservoir and giving reserves when there is a peak demand. Again, this is not a global solution.

The damming of water is also not new but again carries with it many environmental problems from its installation to its risk of possible destruction by natural or other disasters. However, the hydro electric dam is an environmentally friendly resource when running in its normal operation and thus this book will use the good qualities of a dam combined with the power of compressed air to show how a system can be built that is not only environmentally friendly but has the capability of being duplicated anywhere in the world.

Please read on with an open mind.

The Future of Electrical Generation

The Future of Electrical Generation

Compressed Air: its production and power

We are all familiar with the air compressor that people have in their garage for powering tools, paint sprayers and many other uses. They are usually powered by electricity from an outlet or are connected to a gasoline powered internal combustion engine. They are associated with noise as they roar into operation. In factories large containers of compressed air are connected to machinery to supplement the use of electricity in the manufacturing process. Such compressed air is usually brought to the factories by large trucks, carrying a tank of high pressure compressed air to fill the factory's containers. But what makes the use of such compressed air useful, is the psi (pressure per square inch), of such air. Robots are becoming prevalent in many factories, and many are powered by compressed air.

As stated in the introduction, companies that produce compressed air, do so by using electrical or gas engine compressors. However, if high psi pressures are needed a large press like device is used to compress air to a 1000 psi and above. Obviously special tanks need to be built to hold such a pressure and the use of carbon fiber and other such materials meet those needs. In Europe,

The Future of Electrical Generation

there is an automobile company that uses such high-pressure air to run a car and a van and has quite a good mileage record, but its one draw back is that there are not enough refueling stations, at present, to allow owners to refuel.

Historically steam engines were used as the means of producing compressed air but in our modern world, windmills, producing electricity, have been employed to power the compressors. Huntorf, Germany is a good example of this, where such a compressor was built in 1978.

One innovation, that has been making a come back is a system known as a "trompe". It was first used about 1560 to drive a forge in Catalan, Spain, and is also known by the name of the "Catalan Forge". Basically, water from a river is diverted into a funnel like hole, which has many drinking straw-like tubes in it. The river should either have a waterfall nearby or else have a lower level of about fifty feet. The water enters the funnel and as it passes the straws, it "sucks" in air in the form of bubbles. These bubbles travel down the hole using the kinetic energy of the water, and as they do the bubbles are compressed. One draw back to conventional air compressors is that the air they produce is hot due to the friction involved, but in the trompe the air is cooled by the surrounding water and thus the compressed air produced is

The Future of Electrical Generation

isothermal, and thus is not hot, making it a more efficient power source. The "hole" is usually a cylinder that can be 300 – 350 feet long, and as the air reaches the bottom of the shaft, a steel cone separates it and a cavern collects the compressed air, while the water returns to the river via an exit pipe. (A rule of thumb is that for every 25 feet of drop the air is compressed about 10 degrees, although we must take into account the difference between the input and output levels.) Thus, a drop of 300 feet should produce a psi of about 120 psi.

The resultant isothermal air, being cool, means that besides the air quality being better than our atmosphere, the air is a very efficient form of power.

The above method of producing such air was studied and put into practice by Charles Havelock Taylor, when he built a compressor, known as Ragged Chutes, in Cobalt, Ontario, Canada in 1910. It was used to provide fresh air to the miners of the area, while working in the mines, and was the means of powering their pneumatic drills and other equipment. A similar compressor was also built in Michigan, USA, at the Victoria mine.

One feature of the system was called a blow-off or relief pipe, that was used to release the compressed air if it exceeded its intended psi. It worked by having a funnel situated below the

The Future of Electrical Generation

water level in the bottom cavern. As the storage tank filled with compressed air, the pressure was able to push the water level lower, until at a certain pressure, the water was caused to go below the relief pipe. At this point the pressure build up was able to push the column of water in the relief pipe out of the pipe and into the surrounding air, to a height of about one hundred feet. The effect lasted a few minutes and can be seen in You Tube videos, under the title of Ragged Chutes. Although this blow off pipe was intended to relieve air pressure it is the basis for the process that I will describe later and thus I have named it the Taylor ram pump effect.

The trompe is, therefore, a very efficient way of producing compressed air without using mechanical means and the usual related noise associated with air compressors. Its draw back is that to produce a trompe there needs to be a height difference between the entrance to the trompe and the exit. A waterfall or other change in level of a river can provide this difference but it also limits the direct use of a trompe to the immediate area of such a fall and would therefore not be a good source of compressed air for global purposes, except for the fact that compressed air can travel vast distances, in stainless steel piping, with very little loss of pressure.

The Future of Electrical Generation

In the late 1800's a man named Victor Popp managed to construct a compressed air system in Paris, France that extended 30 miles around the city and powered many machines and devices. His system used cast iron pipes and rubber elbows that created an efficiency of between 50 and 67 %, due to many leakages along the way. This figure has been drastically improved upon by using stainless steel piping and connections. Thus Mr. Popp showed the world that the transmission of compressed air power over long distances was and is possible.

Methods of compressed air storage, known as C. A. E. S. (Compressed Air Energy Storage), has been demonstrated in various forms. The most popular is to use large salt caverns so that any moisture in the compressed air is soaked up by the salt, leaving a clean efficient power supply. The operation uses conventional electric generators, used in large generation plants, to provide compressed air to the cavern when the price of electricity is at its lowest, usually at night, and then using the compressed air to turn the same generator when the price of electricity is higher and the demand for electricity is at its peak. (When the compressed air is fed into the electrical generator it is heated by natural gas to make it more efficient.)

The Future of Electrical Generation

Thus, we have seen from the above that compressed air is not only a very useful power source, but that its production, ability to be piped long distances and storage make it a valuable resource. We have also learned from the above that historically although the knowledge of using compressed air was available, the use of coal, oil, natural gas and nuclear power has, up to now, put the large-scale use of compressed air in electrical generation, impossible.

The Future of Electrical Generation

The Future of Electrical Generation

Hydro Electrical Generation using Dams

One method of generating electricity that has been used world wide is that of dams to hold back large volumes of water and then using its kinetic energy to turn turbines connected to electric generators. Some famous ones include the Hoover Dam in the United States of America and the Three Gorges Dam in China. Efficient and reliable are two adjectives to describe their operation but generally there are some drawbacks to using dams.

When a dam is initially planned, there is a need for several streams or rivers to empty into a central area that once dammed, will create a large area of flooding behind the dam wall. This creates both an environmental problem to the local eco-system and the relocation of people living in the flooded area. This may affect the animal population of the area but in most cases the major problem involves the fish that use the rivers or streams for spawning. Some dams have special spillways to allow the fish population to move through the area but in many cases large fish reserves are depleted and sometimes the dam has caused the extinction of some fish species.

Once a dam has been built, the water that enters it eventually becomes stagnant due to algae and

The Future of Electrical Generation

sediment build-up. In normal rivers, the silt or sediment flows with the river and when mixed with the fish feces, becomes a nutrient rich mixture that feeds the local foliage. However, the dam not only stops this natural progression but whereas the river water is usually warm in natural form, it becomes cold and oxygen starved as it leaves the dam.

This silt also builds up behind the dam wall and if not removed periodically can interfere with the operation, or if left unchecked, can eventually stop the dam's ability to generate electricity altogether.

The dam wall construction is also a major concern, besides its usual cost in dollars. Although the base is made of reinforced concrete and is very thick, the volume and weight of the water the dam wall holds back is substantial. Time is never a friend to a dam, and regular inspections of the structure are essential to ensure that there are no cracks or fissures starting. Natural disasters, such as earthquakes, are a threat that can not be taken lightly and the affect of such on a dam wall can be catastrophic, especially to towns or villages in the path of the torrents of water that could be unleashed.

World wide availability of possible sites for future large dams is also limited and thus the future of the dam may be in jeopardy. However, the principle of

The Future of Electrical Generation

using the kinetic energy of falling water is still a viable prospect, so long as the solution does not carry with it any of the above problems.

The Future of Electrical Generation

The Future of Electrical Generation

Rain Water Mounds

One practical way of collecting water at a height is to use rainfall. What is envisaged is shown in Fig. 1 where a mound is created so that the top acts like a giant water barrel. The dome shaped structure is made of rigid plastic that will not crack in the winter and on its perimeter are a series of small holes that allow the rain water to pass into the interior of the storage unit. The structure will probably be built to resemble a geodesic dome as this will support it when snow will cover the structure. The dome will also have a thin layer of concrete to give it further strength, making sure that the perimeter holes are not covered. Also, around the perimeter will be a wall, whose height will equal the height of the dome. This will mean that in heavy rainfall or when the snow melts, the dome will act like a pool to contain the water until it can drain through the perimeter holes.

Inside the dome is another pool that would be ten to twenty feet in depth, allowing the rain water to accumulate. On two sides of the pool will be a series of siphons which will allow the accumulated water to flow into the exit channels when the level of water in the pool covers the top of the siphons. This now creates a flow of water that will be

The Future of Electrical Generation

channeled into a trough containing several plastic tubes that will act as straws to draw air into the flowing water and thus give plenty of air bubbles. The two channels will then meet in the middle of the space below the accumulation pool and enter an oblong pipe that will extend 300 feet into the ground. (Fig. 2) At the bottom of this shaft will be a stainless-steel separator that will force the now compressed air bubbles to rise to the surface of the water and enter a large cavern that will have a dome to collect the compressed air. Another pipe will then direct the water up 250 feet to enter a second rain water mound that is fifty feet lower than the first mound. Fig. 2 also shows a funnel shaped pipe that will act like a relief valve when the air pressure in the cavern builds to a greater psi than is needed, (much like the one described in the Charles Havelock Taylor compressor in Ragged Chutes), but instead of the water going back into a river it will travel up the relief pipe to exit above the mound and thus enter again through the dome into the accumulation pool. Thus, although most of the water will return to the system, some will be lost due to evaporation.

The compressed air (about 120 psi) will be piped into a storage gasometer close to the air electric generation plant.

The Future of Electrical Generation

Rain water mound one will therefore be built to be 150 feet above ground so that when the exit water enters rain water mound two, it will enter the accumulation tank of mound two, and thus mound two will be approximately fifty feet lower than mound one. However, the oblong pipe described above, resembling the one in mound one, will also extend 300 feet and thus produce compressed air of about 120 psi (Fig. 3).

Rain water mound three will therefore be built fifty feet lower than mound two and when the water exits mound three it will be at ground level and can therefore be channeled to an existing river or stream nearby. Thus, the rain cycle has not been disturbed but the rain water will have been used to produce large amounts of compressed air at about 120 psi.

It was mentioned earlier that compressed air can travel long distances with very little loss of pressure, which means that these rain water mounds do not have to be too close to the actual air electric generation plants. Thus, areas where there are frequent rainfalls would be best for the mounds. Such locations would be close to lakes, rivers or streams. Obviously, areas of the world where there are cold winter climates would restrict the use of such mounds, although as mentioned, the domes can be covered with snow until the

The Future of Electrical Generation

spring thaw and thus they could operate in the other three seasons of the year. It would also mean in such areas, gasometers to store compressed air would be constructed, probably below ground, to handle the winter demand for the compressed air.

Some other ideas of using different methods of raising water are expressed in another book of mine, entitled "Suffer the Little Children".

Note that the rain water mounds are only used to produce compressed air and will not act as electrical generators. It could also be a consideration in such areas that are in cold climates, that alternative energy such as wind, solar and wave power could be used to drive a conventional air-compressor, or a circulating system could be constructed that would be insulated against the cold and thus allow the water to flow through the trompe to produce compressed air. The decision would therefore be made when it is determined how much compressed air will be needed to run the air electric generation plant during the winter months.

The Future of Electrical Generation

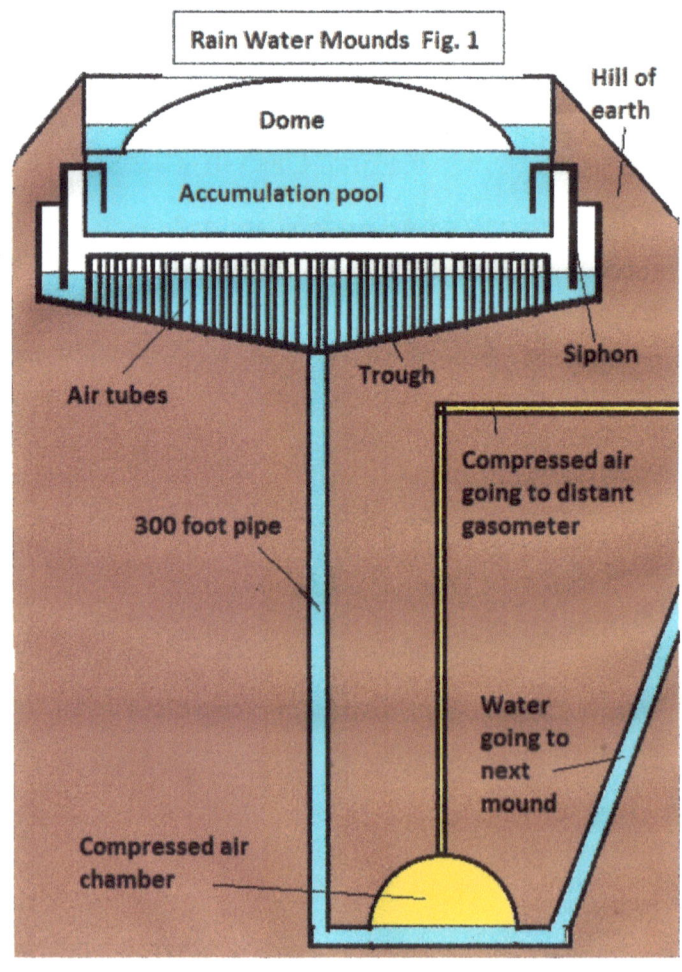

The Future of Electrical Generation

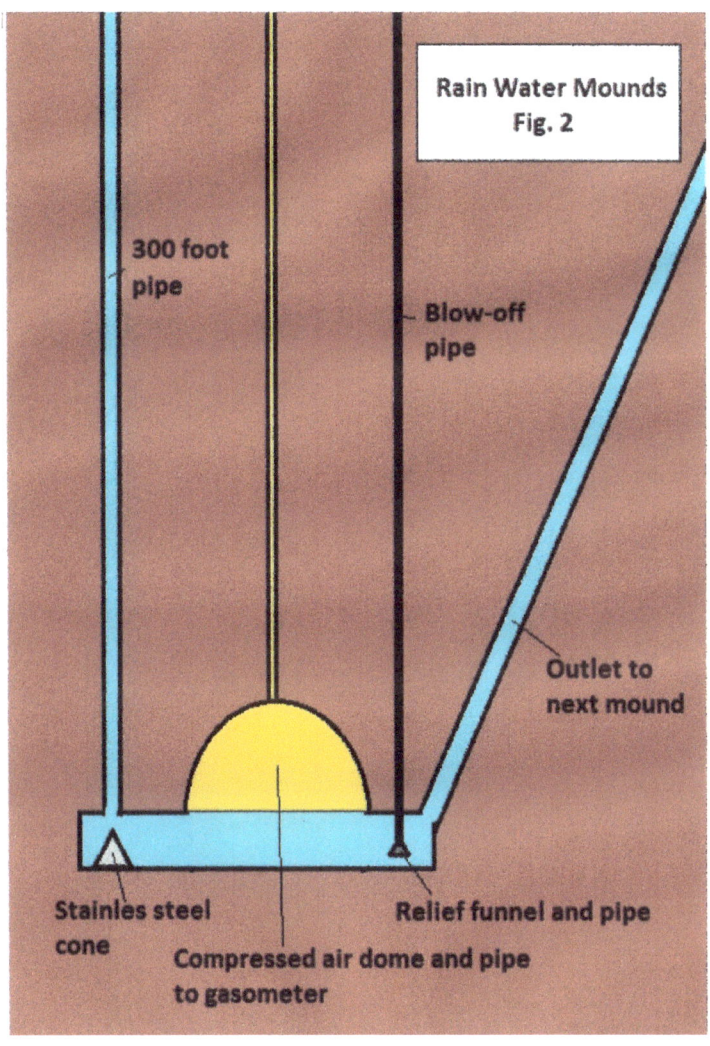

The Future of Electrical Generation

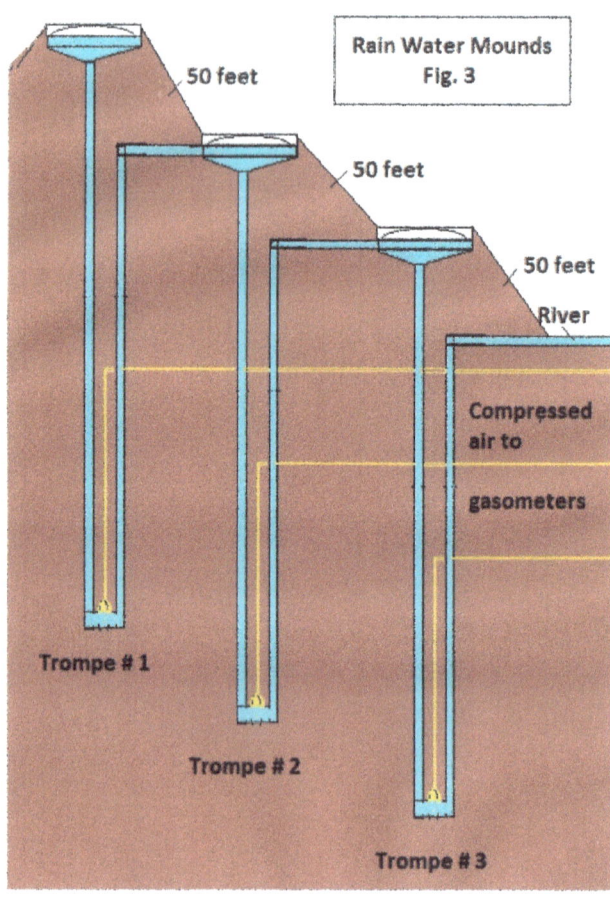

The Future of Electrical Generation

Taylor Ram Pump and its Adaption

When the Charles Havelock Taylor compressed air trompe was described earlier, mention was made of the relief pipe delivering a burst of water and compressed air when the air pressure in the cavern exceeded the required psi and the water had been pushed below the level of the funnel connected to the relief pipe. For this reason, I have called this phenomenon the Taylor Ram Pump although its adaption is not quite as envisaged by Taylor. Fig. 1 shows a bottle that is filled with water. From its lower side, an aquarium air pump is attached, with a one-way check valve between the pump and the bottle. A little above this connection is a second pipe that extends four feet above the bottle and empties into a jug. After the bottle is filled with water the bottle cap is screwed back onto the bottle. Note that when the bottle is filled, the water will also enter the four-foot pipe but only rise to the same level as the water in the bottle.

The aquarium pump is turned on and immediately the air from the pump, bubbles to the top of the bottle starting an increase of air pressure in the bottle. This causes the water in the exit tube to rise and eventually the water reaches the top of the exit tube and empties into the jug, (Fig. 2). As the

The Future of Electrical Generation

air from the aquarium pump continues to fill the bottle, the water continues up the exit pipe and into the jug. (Fig.3) When the air in the bottle reaches the exit pipe level the whole compressed air in the bottle pushes what water is in the exit tube, and as with the Taylor compressor, this water shoots out of the exit tube. Thus, now the bottle contains ordinary pressure air and remains so when the aquarium pump is also turned off. When the cap is taken from the bottle the air pressure is normal (15 psi) and therefore if water is above the cap it will readily enter the bottle, (Fig. 4).

When the bottle is filled again with water and the aquarium pump is turned on, the same cycle as above will begin. This, then, is the principle behind the adaption I call the Taylor Ram Pump.

Taking the above a step further, Fig. 5 shows a unit that has three compartments, two are three feet deep, three feet wide and three feet long, one is four feet deep, three feet wide and three feet long. The compressed air is pumped into the first foot of water in Tank B and there is a one-way valve between Tank B and Tank A. Six exit pipes are not shown but extend to Tank C. Against Tank A is Tank D which empties into Tank A. From Tank C there is a siphon pipe that has a one-foot short side in Tank C and extends six feet to a turbine connected to an

The Future of Electrical Generation

electrical generator. There a short pipe from the turbine that takes the water to Tank D.

To begin the operation of this demonstration, Tank D, A and B are filled with water and the compressed air supply is turned on so that about forty psi is pumped into Tank B. The one-way valve is closed. The air pressure will cause the water in Tank B to enter the exit pipes and rise the twenty feet to Tank C, until the air in Tank B reaches the exit pipes and shoots the remaining water into Tank C, (Fig. 6).

If we now, using the above idea, expand it to increase its capacity, a system can now be shown to create as it were, an elevator of water that can raise a volume of water to great heights. Fig. 7 and 8 shows how this will work.

Each unit is shown to have a depth of three feet with a length of twelve feet and a width of twelve feet, thus creating a space of 432 cubic feet. This space, if filled with water, will hold 3,231.58 US liquid gallons. In tank A are three check valves that will be described in the next section. Below Tank A is another tank, shown as Tank B, which is four feet deep, with a length of twelve feet and a width of twelve feet. On one side is an array of pipes that are ½ inch inner diameter that are spaced ½ inch apart, thus there will be 144 pipes. Each pipe is twenty feet long and extends to the next unit

The Future of Electrical Generation

above it. Each pipe will enter Tank B at the one-foot mark, thus the depth in Tank B where the pipes exit, will be three feet. Below that three feet is one foot of water in which compressed air is pumped through two pipes. As with the bottle example, shown above, the compressed air will be pumped in until the water in the exit pipes have received their burst of air that will force both the air and water in the pipes into the next level (Tank C). At that point the check valve for the compressed air will stop the flow and the check valve for the water will open allowing the contents of Tank A to enter Tank B. Thus, the contents of Tank B, (at the three feet depth), will pass from Tank A into Tank B and onto Tank C, creating the elevator effect.

It is envisaged that there will be twelve units, each 12 ½ feet high, so that the elevator effect will go to 150 feet. Thus, the 3,231.58 gallons of Tank A will eventually reach the top of the elevator (Tank D) and then enter Tank E, (see Fig. 9). Tank E will be nine feet deep, twelve feet wide and twelve feet long, so that the siphons in Tank E will empty a third of its contents before the next input of water from Tank D will fill Tank E again.

The water from Tank E will then cascade down the siphons for one hundred feet, where the kinetic energy will enter a Tesla turbine connected to an

The Future of Electrical Generation

electrical generator. The generator will be the conventional one used in hydro electric power stations presently. The run off from the Tesla turbine will then enter a channel which slopes slightly so that the water can pick up air bubbles as it passes under several thin plastic tubes and then falls into a trompe that will extend at least 300 feet below and like the rain water mound, will have stainless-steel plates that will separate the air from the water. The cavern will have a compressed air compartment that will lead to a gasometer where compressed air from the rain water mounds will also be stored. This gasometer will then feed the several pipes that will go to the water elevator previously described. (Note that this is the point where compressed air from the rain water mounds will combine with the compressed air of this trompe.)

Thus, the water will again begin at the bottom of the elevator and the cycle will continue, generating electricity as it goes. Since the elevators of the air electric system are independent of one another, there can be several units joined together, so that as electricity demands rise and fall, each unit can be stopped or started at will by stopping the check valves for water and compressed air.

The Future of Electrical Generation

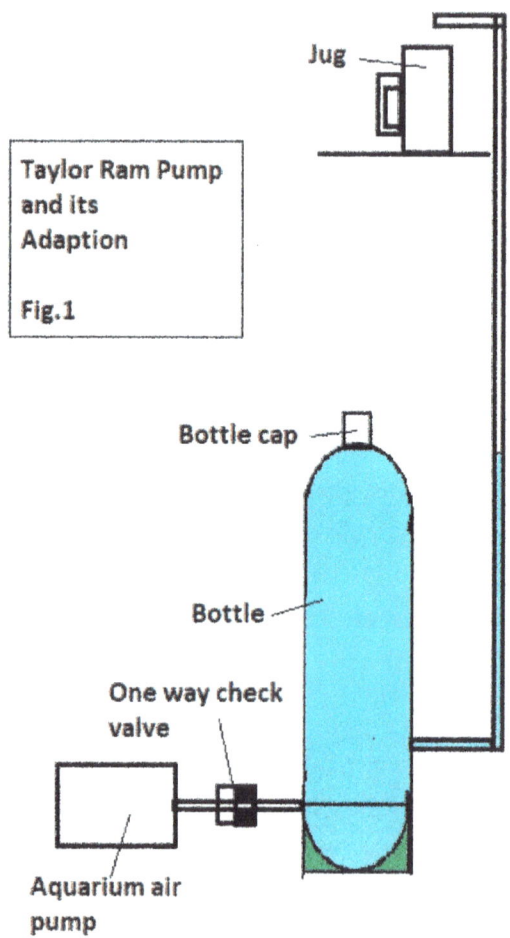

Taylor Ram Pump and its Adaption

Fig.1

The Future of Electrical Generation

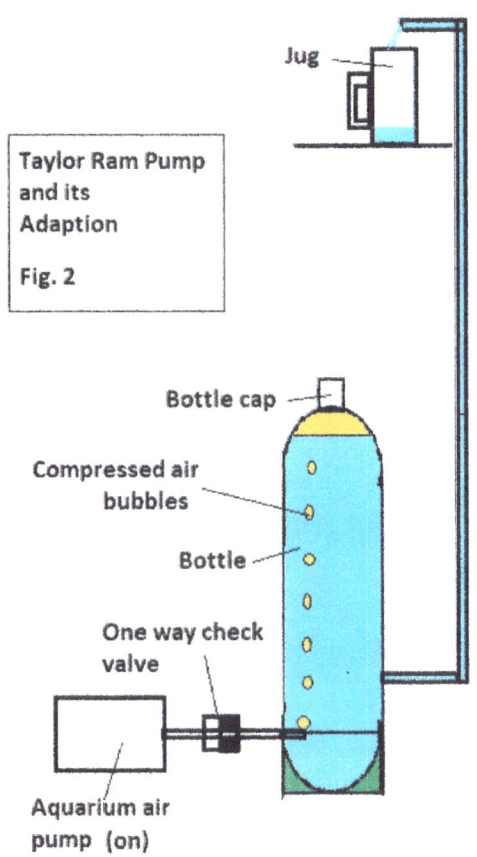

Taylor Ram Pump and its Adaption

Fig. 2

The Future of Electrical Generation

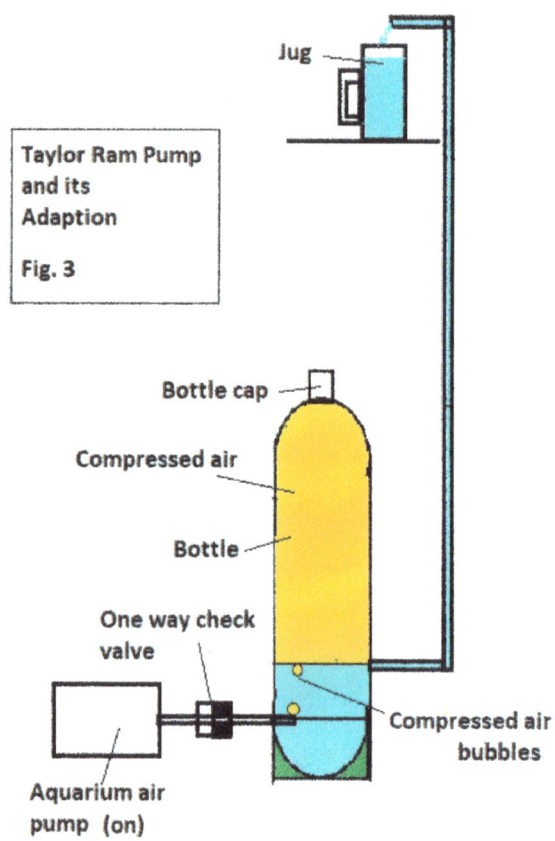

The Future of Electrical Generation

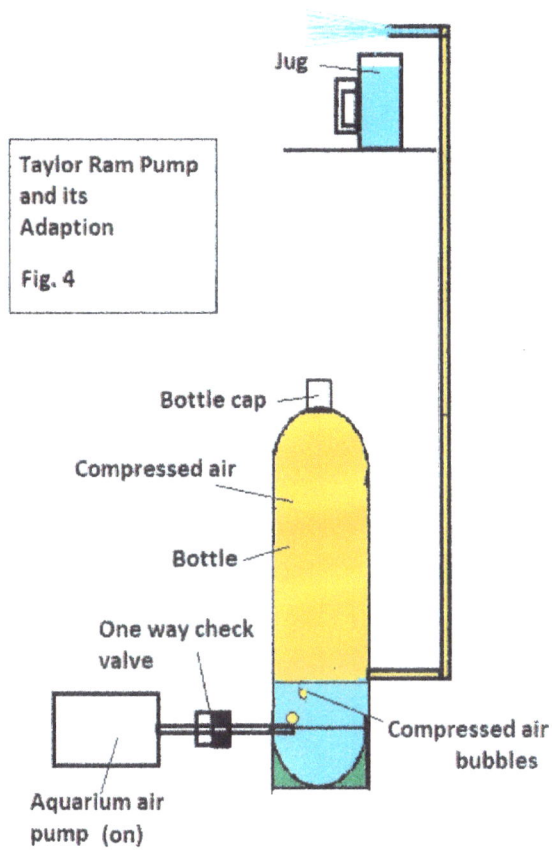

The Future of Electrical Generation

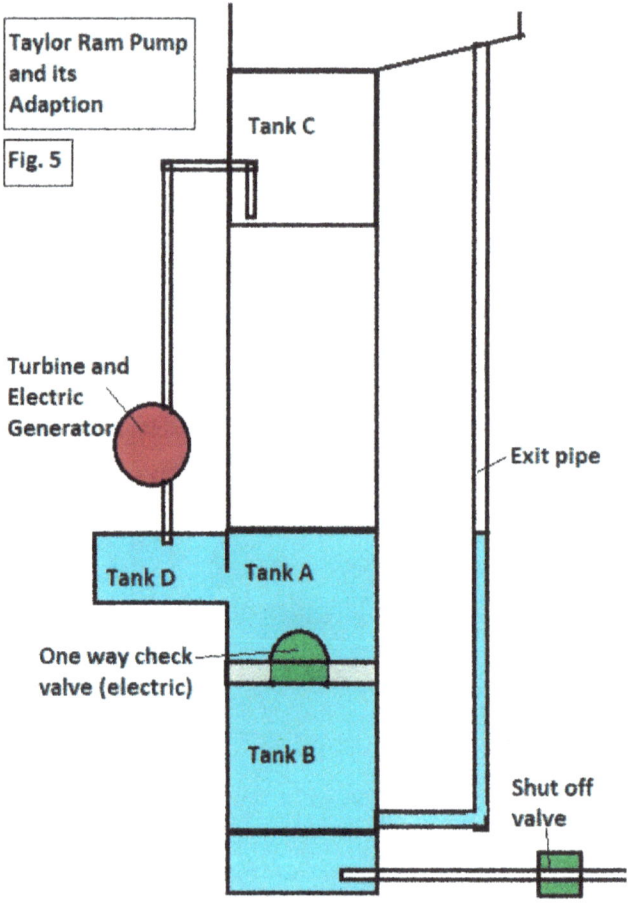

The Future of Electrical Generation

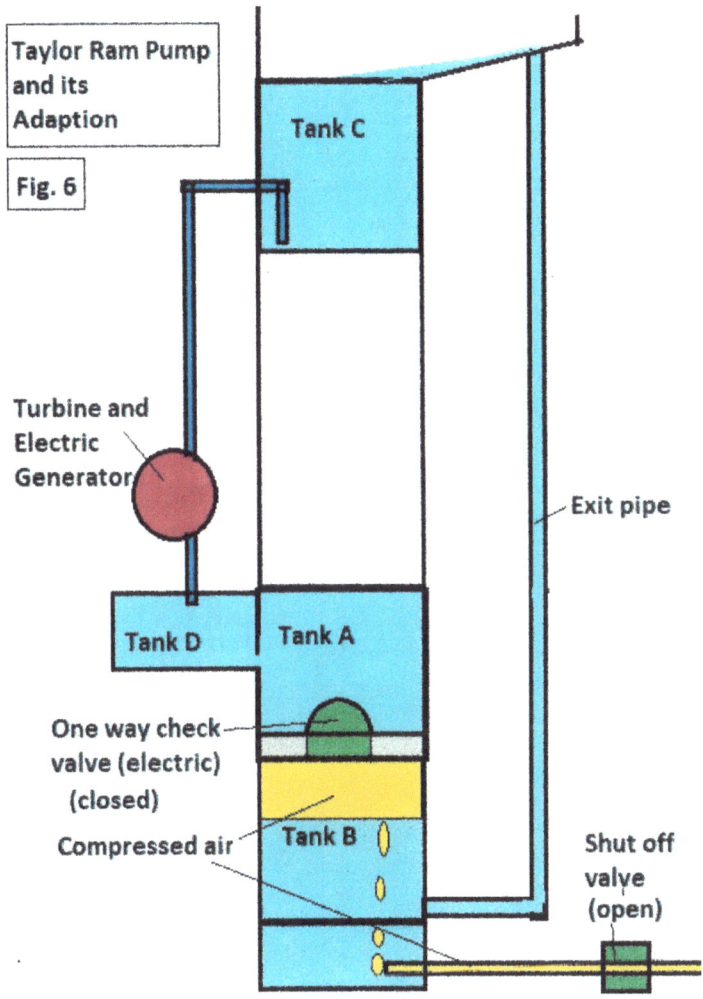

The Future of Electrical Generation

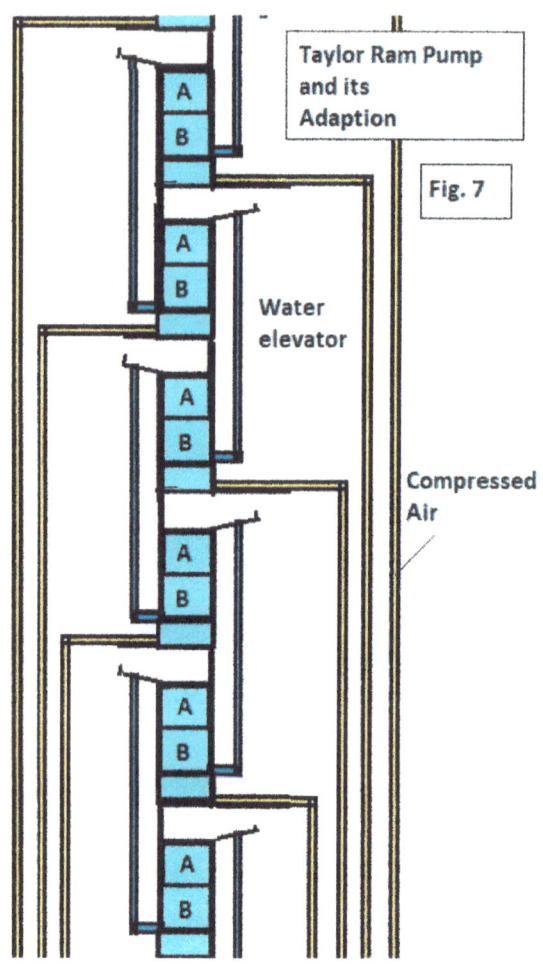

The Future of Electrical Generation

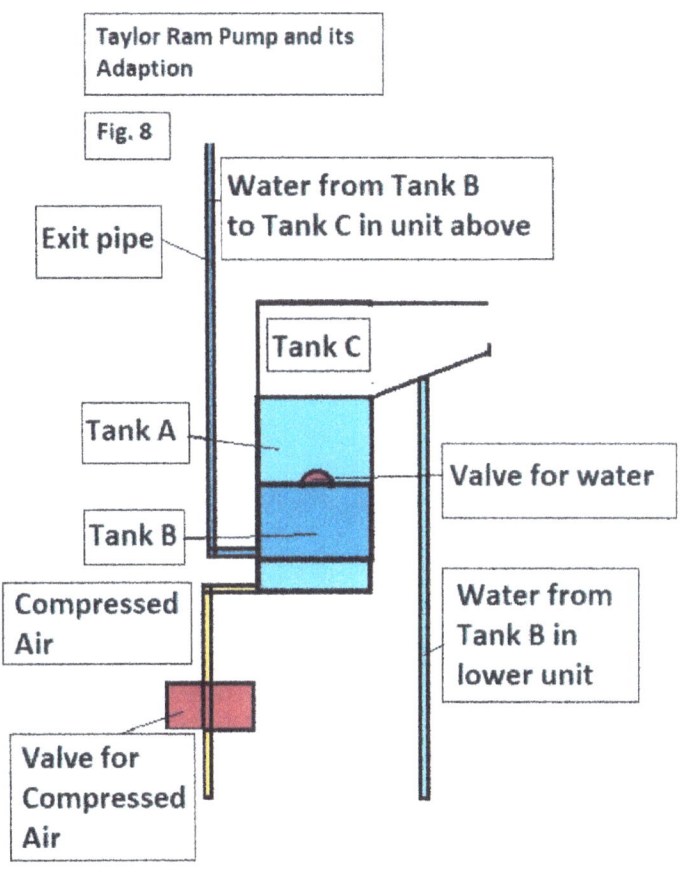

The Future of Electrical Generation

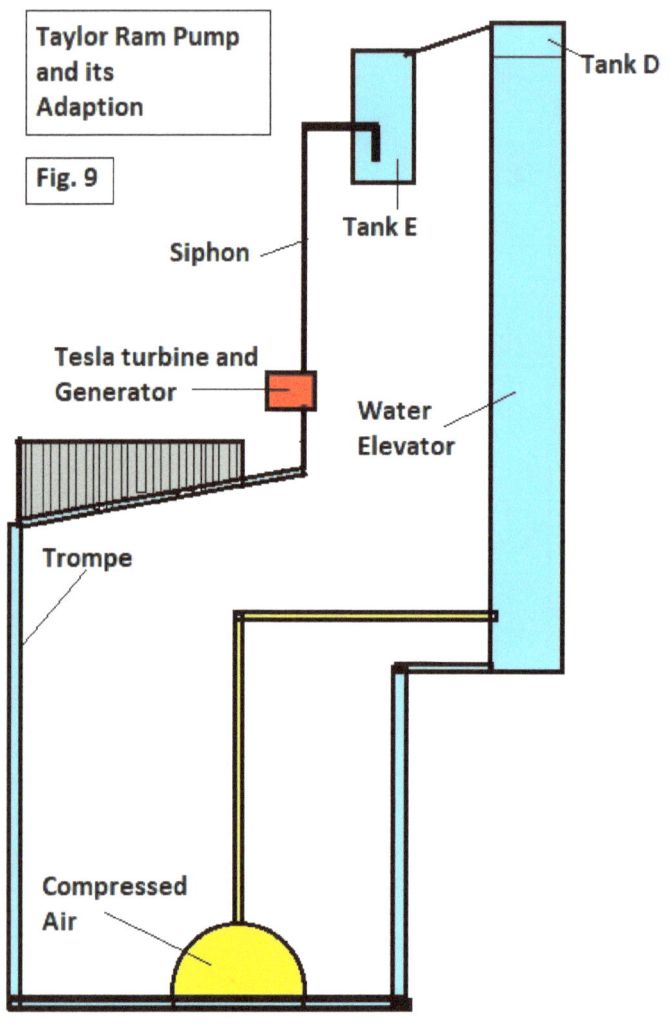

The Future of Electrical Generation

Check Valves for Water and Compressed Air

When I described the air electric system it was mentioned that there would be check valves to allow water to move from Tank A to Tank B. In this section I will show a valve that I feel would best suit the task. Fig. 1 and Fig. 2 show how the valve will work. The action of causing the ball shaped stopper to move up and seal the hole where the water enters, and down to allow the water to enter, is created by using both permanent and electro-magnetism. The reason for this is that once the electro-magnetism is activated the permanent magnet will be repelled in one direction and attracted in the other, but once the permanent magnet has moved it will stay in position and thus the electro-magnetism is activated for only a short time.

When the valve is in the closed position and the stopper is pressing against the entrance to the valve, the compressed air entering the tank will begin to build and thus, the stopper will be further pressed against the entrance. This action will continue until the compressed air leaves the tank via the exit pipes. As mentioned earlier, the air pressure in the tank will then return to normal so

The Future of Electrical Generation

that when the valve is activated to the open position, the water pressure above will aid in moving the stopper away from the entrance.

Since the elevator effect will occur in each tank unit at the same time, the activation switch will only have to be in the first unit and it will then be transmitted to all the units above. Thus, when unit one is allowing water to flow from Tank A to Tank B, the same action is happening in all units above. The compressed air valve will also open when the water valve closes and the action will also be repeated in each unit at the same time.

It can therefore be seen that the activation switch for both the water and the compressed air, can be controlled in the first unit.

Fig. 3 and Fig. 4 show how the compressed air control will work. Although this valve uses the same principle of using a permanent magnet and electro-magnets for movement, the stopper is a cone shaped rubber pointer instead of a ball. The pipe carrying the compressed air enters Tank B where the permanent and electro-magnets are in the tank. When the air is stopped the pointer is inserted into the mouth of the compressed air pipe but when the pointer is withdrawn, it allows the compressed air to escape through the water in Tank B, as air bubbles.

The Future of Electrical Generation

The check valves for both the water and the compressed air are simple in design but effective in their operation, which is important since the elevator cycle will demand a constant flow of water. The electricity to activate the valves will be provided by batteries, which may be lithium ion or zinc bromide, but their source of energy will come from either wind (windmills) or solar (solar panels) or a combination of both. As explained above, since the electricity will be minimal for each stroke of the valve as the permanent magnet will sustain the action of keeping the valve either open or closed. (The electricity may also come from the generated power of the turbines, run by the water elevator.)

Since both the water check valve and the compressed air check valve will be under-water, they will be protected by being encased in plastic, however, such plastic will be thin enough to keep the water out but not effect the action of the magnetic attraction or repulsion.

The Future of Electrical Generation

Check Valve for Water and Compressed Air

Fig. 1

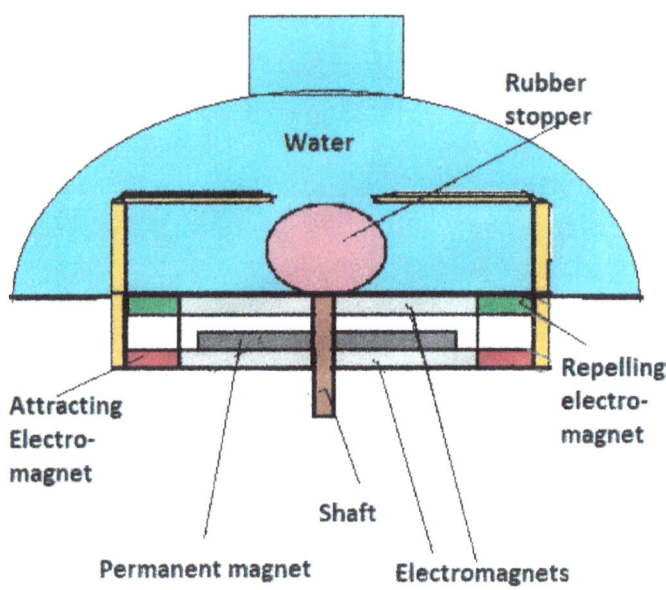

The Future of Electrical Generation

Check Valve for Water and Compressed Air

Fig. 2

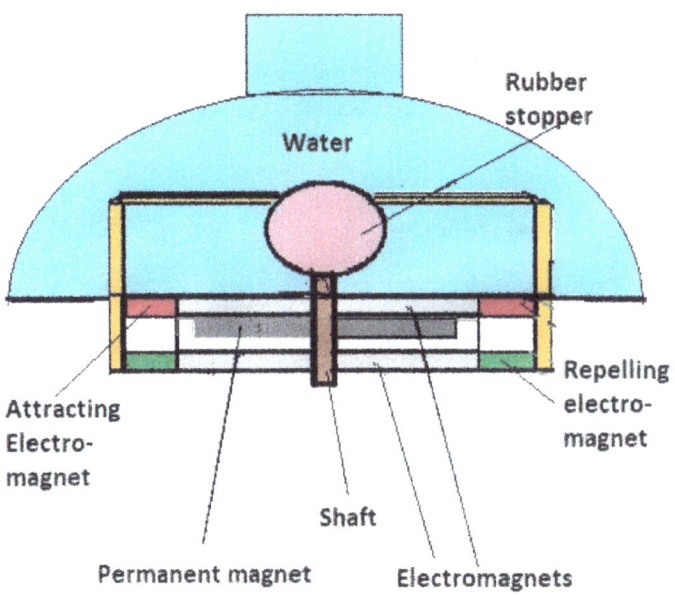

The Future of Electrical Generation

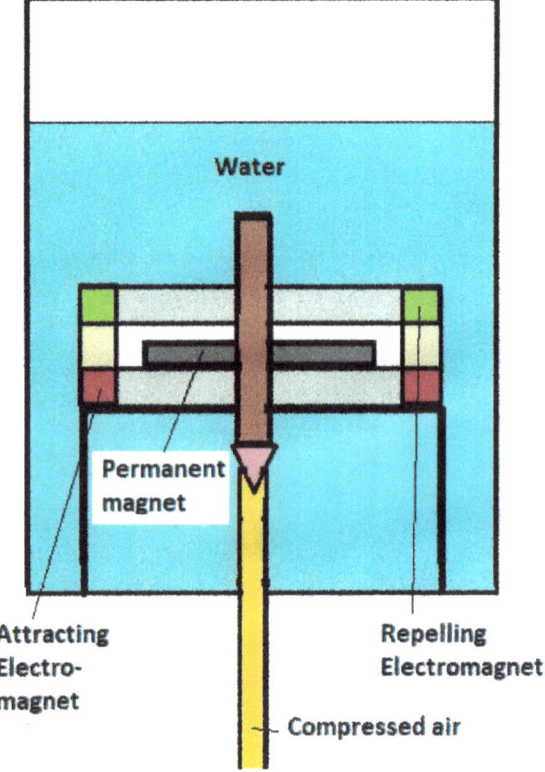

The Future of Electrical Generation

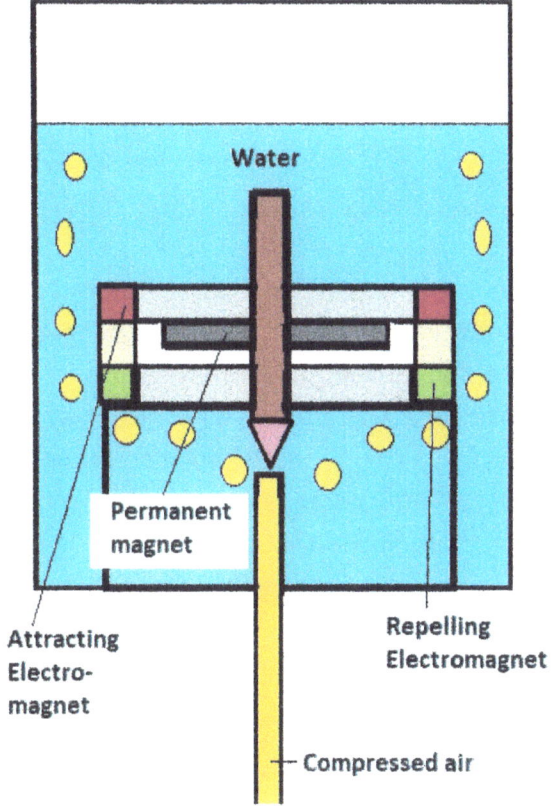

The Future of Electrical Generation

Volumes of Water Involved

Earlier we mentioned that the air electric system would be made up of several units that would be divided into sections or Tanks. Tank A would be three feet deep, twelve feet wide and twelve feet long, giving a cubic volume of 432 cubic feet. Tank B would be four feet deep, twelve feet wide and twelve feet long giving a volume of 576 cubic feet. (Tank C acts as a reservoir and does not need to be in the equation.) The water in the exit pipes would account for another 18 cubic feet, thus, the total cubic feet would be 432 + 576 + 18 feet or 1,026 cubic feet. Since the air electric system will consist of twelve units high to produce an elevator 150 feet high, the total cubic feet of water will be 12,312 cubic feet. When that is equated; it will be 92,100 US liquid gallons, which would weigh 768,612 pounds or 384.31 US tons.

Thus, the structure built would have to support the 384 US tons of water plus the weight of the twelve units. At the top of the elevator would be the tank that siphons the water to the turbine, one hundred feet below. The weight of the water in the upper tank will be a volume of twelve feet by twelve feet by twelve feet, or 1,728 cubic feet, or 12,926.34 US gallon, or 107,875.34 pounds. Obviously, that will

The Future of Electrical Generation

be a lot of water going down the siphon tube to drive the electric-generator, but the diameter of the siphon pipe will be determined by the amount of water that is extracted from the twelve-foot tank and how quickly the water elevator is able to refill the water taken. One special feature of a siphon is that when working, the volume of water in the pipe will be consistent and thus its output will also be consistent, and since our aim is to produce electrical power at 60 cycles per second (cps) in America and 50 cps in Europe, such consistency will allow us to do so.

The amount of time that the units of the water elevator will take to refill will be controlled by the amount of air pressure that is applied to Tank B. We have mentioned earlier that the trompe will be able to produce compressed air at about 120 psi. This may be too much for the air electric system, and so experimentation will provide the correct air pressure requirement, which is obtained by the depth of the trompe. We have mentioned that a trompe 300 feet deep will be able to produce about 120 psi, so if a lesser psi is needed then the trompe depth will be shorter and vice versa.

So, as with any fine-tuned machine, once the pressures, depths, timing and siphons have been calculated, the air electric system can begin. Once the operation begins, the output of electrical

The Future of Electrical Generation

power will be consistent and, as mentioned earlier, the determined output will be able to be ramped up simply by starting new water elevators.

The one advantage of this system is that as the water is constantly oxygenated and circulated, there is little chance of algae or other problematic organisms affecting the operation.

The Future of Electrical Generation

Tesla Turbine and Generator

The Tesla turbine was described by its inventor as his favorite invention, in that Nikola Tesla felt that this turbine would revolutionize the way the world would use turbines. However, when he invented it there was not the proper materials to build it so that the turbines could take the stress of the high speeds that are achievable. Carbon fiber, graphene and other materials being discovered should make the production of such turbines possible.

The turbine uses the Bernoulli principle to guide water that adheres to its surfaces and spins the turbine, taking a vortex path to the exit holes. (The Tesla turbine is more fully described in my other books, "Can Tesla Save the World?" and "Suffer the Little Children".) You Tube also has many videos to describe the operation of the Tesla turbine. Its main advantage over conventional turbines, such as the Francis turbine, is that its efficiency is such that Nikola Tesla boasted that it could be as high as 98%.

The other fact that gives this turbine its appeal is that its operation is very simple and does not require the many intricate fins of a conventional turbine, which also means that it will require lower

The Future of Electrical Generation

maintenance. Although the Tesla turbine can work using compressed air, steam and water, for our purposes we are only interested in water.

In Fig. 1 the Tesla turbine is shown as being horizontal instead of the usual vertical position, the reason for which is that when the water goes through the turbine it will exit below and thus not impact the generator that will be above it. Note that there is a small umbrella on the connecting rod, to also help keep the water out of the generator.

The electric generator can produce alternating power, (as mentioned before at 60 cps in America and 50 cps in Europe), by rotating electro-magnets near to fixed electro-magnets. The wattage produced by each electric elevator will depend on the diameter of the siphon used but it is expected that many megawatts will be achievable. As also mentioned earlier, there is the capability of starting and stopping various water elevators to meet the demand needed for the grid or other use of the generators.

The Future of Electrical Generation

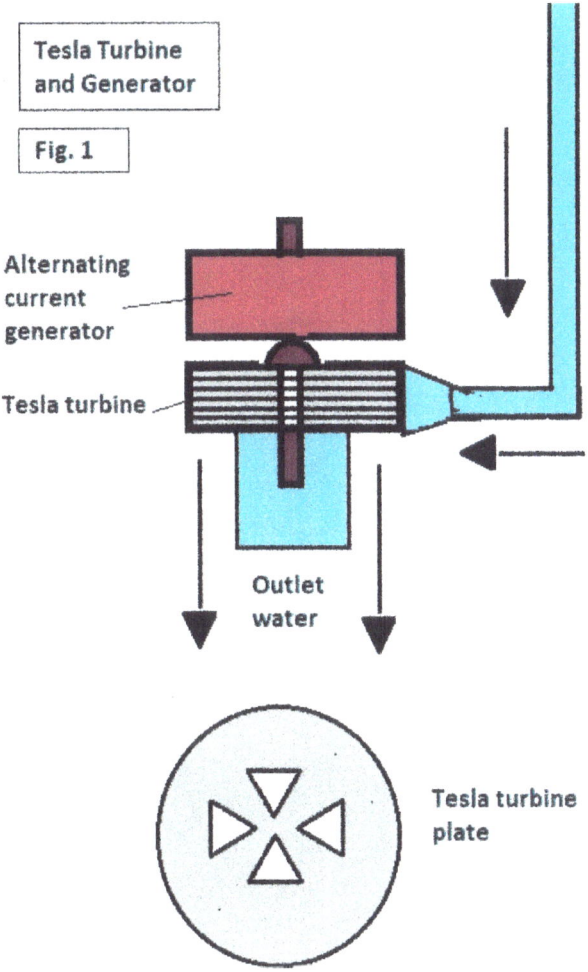

The Future of Electrical Generation

Production of Distilled Water

Although the air electric system can run on ordinary rain, lake, river or stream water, ideally distilled water would be preferable for its many advantages. However, the amount of distilled water initially used would seem substantial, but it is envisaged that after several air electric systems are in operation, one can be used to produce steam that would be condensed to make the distilled water. One of the advantages of such water is that as it does not have contaminates in it, the chances of having any bacteria growing is eliminated and since the water will be constantly recycled within the system, there is no need for any other water.

The compressed air that will run the water elevator may introduce some contaminates but a little chlorine periodically will take care of that. Such compressed air is also a way of oxygenation and will allow the water, being that it is constantly circulating, to maintain its purity.

The distilled water will also provide another source of income, if bottled and sold to the retail sector.

The Future of Electrical Generation

Siphon System for Transporting Water

In a previous book of mine, (Suffer the Little Children), I described a way of transporting water on a level surface. The advantages to this are that it is not necessary to have a gradient for the water to flow downwards, however, although the water can flow on a level plain, a gradient is also desirable.

The water for the siphon system begins in a reservoir that may be ten feet deep, into which is a pipe that is three feet long. The water then flows on the level through a pipe to the next reservoir, where the pipe curves to a depth of six feet. This head of three feet now provides the basis of a siphon that will draw water from reservoir one into reservoir two. When the water in reservoir one and the water in reservoir two reach the same level, the siphon effect will stop. If a third reservoir is added to the system, the water will now flow from reservoir one to reservoir two to reservoir three, dependant on the water in reservoir one is being fed so that its level is above the siphon. The reservoirs may be extended to any number, again dependant on reservoir one remaining full. The depth of the reservoir is also not important, the

The Future of Electrical Generation

siphon will work if the reservoir is over six feet deep in the above example.

Should there be a gradient over six feet, the siphon can then enter a new reservoir at a lower level and then this reservoir will become reservoir one for a new section. Fig. 1 illustrates the above scenario.

The Future of Electrical Generation

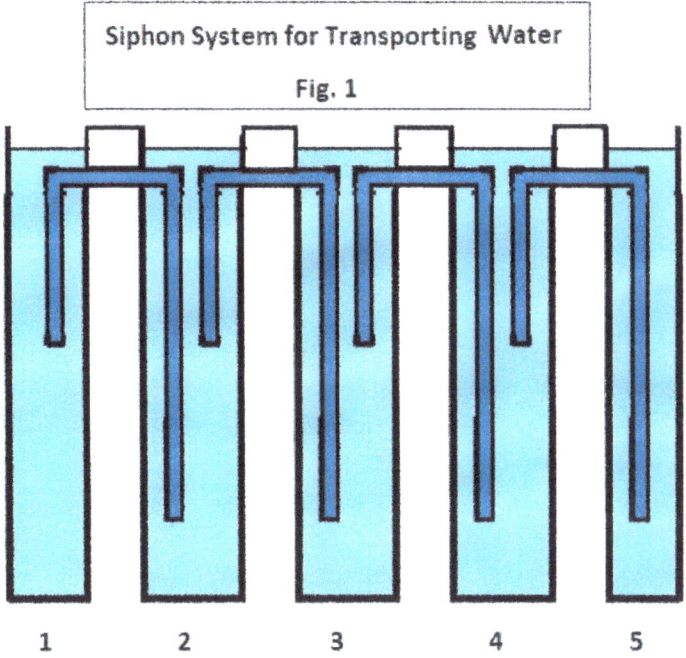

Siphon System for Transporting Water
Fig. 1

Reservoir 1 feeds reservoir 2, etc. As long as reservoir 1 is full the other reservoirs will fill. The distance between reservoirs also does not matter as long as the top of them are level and the siphons are level.

The Future of Electrical Generation

Air Electric System in Operation

The air electric system is intended to eventually surpass all systems that use fossil fuels to generate electricity, but it is realized that political resistance will challenge its progress. Eventually it is hoped that overwhelming evidence of climate change will convince the world that unless this earth's resources are used sensibly, humanity is going to pay a heavy price for its stubbornness.

Since our planet is made up of over 70 per percent water, of which billions of tons of it fall back to earth as rain, it would appear to be the logical answer to our power source accompanied with the use of air, that is, every where around us. The beauty of the air electric system is that it does not change, destroy or pollute its two main elements; air and water. Instead it recycles these elements, improving both as it does. As mentioned before the compressed air produced by the air electric system will be isothermal and thus of superior quality, compared with most of the air we breathe. The water, being oxygenated and constantly moving will be less of a maintenance problem than other water systems and as the valves that cause the system to work use minimum electricity to operate,

The Future of Electrical Generation

the efficiency of the overall production will be very high.

Since the water elevator can be virtually built anywhere it makes sense that if the electricity generated is needed in a town or city, the air electric system can be on the outskirts of that town or city, thus, making the need for power lines and towers traversing the country-side obsolete. The water for the system can be either siphoned or trucked to the site initially and then once the system is full, the operation will be self-sufficient, except for some water loss due to evaporation. Obviously, there may be some fear that such a volume of water near to a town or city may cause a flood, but once those concerns are addressed by showing that the construction of the elevator will be strong and firm and when more than one system has been in service for a length of time, such concerns will go away. It is envisaged that the system will be either integrated into a bluff or local hill but if none is available, an artificial hill or mound will be created from local soil. This will be designed to blend into the local landscape and being covered with grass and possibly bushes will create an ideal picnic area for local-residents. The soil will act as both an insulator and a sound barrier for the running water.

The Future of Electrical Generation

As the air electric system will be close to the town or city, only local transformers and wiring associated with them will be required to feed the homes and businesses with the necessary power. This means that disruptions to the service will be localized so that ice storms and other such disasters will also be localized making maintenance of the system easier to handle. The compressed air to run the system will also be local as the use of gasometers, hidden underground, will be used. The compressed air, gathered from rain fall mounds, can be great distances away and yet be fed to the air electric systems via stainless steel piping, underground.

It has been mentioned that the air electric system can be adapted to any area in the world, which means that there would have to be many variations for the system to meet the requirements of cold areas, hot areas, dry areas, wet areas, high areas and low areas. Although the system will work most efficiently with distilled water as its source, it is versatile enough to be able to use rain water, lake or river water or even sea water. Methods of creating compressed air, besides those discussed in this book, are covered in my other books (How Tesla can Save the World? and Suffer the Little Children), which means that world wide there are many ways to compress air but generally the trompe is the best known.

The Future of Electrical Generation

The Future of Electrical Generation

Advantages of the Air Electric System over a Hydroelectric Dam

Hydroelectric dams have been in existence from the discovery of alternating electricity, when man used the energy of Niagara Falls, and then built such large dams as the Hoover Dam and the Three Gorges Dam. Thousands of such dams have been constructed all over the world, some large, some small, to generate electricity for local and metropolitan areas. Some dams are used for recreational activities, some hold the local water supply for drinking water, some dams create irrigation for farmers and some dams are built as flood control, waste management and wildlife habitats. The dams of this world have therefore met many needs that help humanity to survive. However, for all their advantages there are many disadvantages to hydroelectric dams.

One disadvantage is their initial construction, in that there is usually a need for large areas to be flooded to create the dam. In many cases this means that local-residents need to be relocated, vegetation is flooded, and wild life is bewildered by losing their environment. One species of wild life affected by dams is the local fish supply. Dams can affect their spawning habits, contaminate the

The Future of Electrical Generation

rivers connected to the dam, and cause relative differences in the temperature in that the result of the dam water entering the river is cold, stagnant water.

The cost of building dams has grown to many billions of dollars and the length of time for constructing them can go into decades before the dam is ready for use. The dam must pass a stringent inspection before it can open, which is vital in that if there is the slightest flaw in construction it can have devastating consequences. The inspection can not account for all the scenarios that may occur in the life of the dam, but they try to simulate any hazard that may happen. Earthquakes can be disastrous to a dam, and although sites are chosen in areas where seismic activity is rare, in the changing climatic conditions, we are seeing earthquakes occurring in areas that were formerly considered safe. Heavy rainfalls are also becoming a common occurrence and thus flooding is becoming a norm for many areas. If this flooding occurs in rivers and streams feeding a dam, this also can cause devastation if the water level in the dam rises above the dam wall. Ice can also be a dangerous factor as water expands in icy conditions, putting a further strain on the dam wall.

The Future of Electrical Generation

The reason that water in the dams become stagnant is that silt or sediment that usually is swept along by rivers, starts to build up at the dam wall and if not removed can eventually impede the operation of the dam's ability to generate electricity. This sediment also creates a breeding ground for growing algae and weeds that causes the water to stagnate, and thus effect rivers down stream of the dam; producing cold, stagnant water that affects the fish population.

Having covered many areas of the advantages and disadvantages of a dam, we shall now compare the air electric system. Whereas the dam can only be built in certain areas, the air electric system can be built anywhere, including close-proximity to the need for its electricity, such as on the outskirts of a town or city. The idea of having a hill or mound close to an area can only add to the beauty of the landscape as well as opening a new recreational area for the town or city. The hill or mound also means that whereas a dam is open to the elements, the insulation of the earth creating the hill or mound, allows the water within it to be able to flow all year round with little evaporation to the system.

The air electric system uses a water elevator to raise its motive power to greater heights and thus, unlike a dam, holding back millions or billions of

The Future of Electrical Generation

gallons of water, this elevator creates enough head to run the system, without any unnecessary waste. The water will not be stagnant or full of sediment but be oxygenated and moving so that its quality is pristine. If this water is also distilled, its quality will be further enhanced and as the water elevator is able to recirculate its supply, the water will be easily purified if any contaminates creep in.

The dam uses systems that regulate the supply of water and make the generators supply alternating electricity at precisely 60cps in America, (or 50 cps in Europe). These systems are expensive to maintain and are dependant on the demand of the grid as to what its output will be at any one time. The air electric system can be connected to the grid, but it is intended to supply local needs and therefore is easier to maintain because of local requirements. The use of siphons within the system also mean that exact quantities of water are fed to the turbines, making it easier to maintain the required cycles per second. The Francis turbine is the usual turbine of choice for most large electrical generation stations, but the air electric system will use Tesla turbines that are again easier to control and maintain accurate speeds, as well as being low maintenance.

The water level in the dam may vary due to many factors such as evaporation, flooding of rivers that

The Future of Electrical Generation

feed the dam, rain or snowfall, and the demand for the water by the electrical needs. In the air electric system, the main water elevators are shielded from most forms of evaporation, rainfall, etc. and thus, the levels of water are consistent, but to meet the demands for electricity just means that one elevator is stopped or started, which is a flawless, easy matter to do.

The motive power in both the dam and the air electric system is water and its kinetic energy produced by its volume to turn turbines. However, the air electric system relies upon compressed air to maintain its operation. This compressed air is produced within the water elevator plant, but it is also supplemented by extra compressed air supplied by rain water mounds, waterfalls, or other methods of raising water, but their common factor is that the trompe is used to create that compressed air needed to run the operation. As mentioned earlier, compressed air can travel vast distances with little loss of pressure and therefore it can provide an unending power supply.

Finally, the major advantage of the air electric system is that it is one hundred percent environmentally friendly and as also shown, does not add anything to the atmosphere that is harmful. The dam would also appear to be environmentally friendly but the build up of algae

The Future of Electrical Generation

and weeds, over time, creates methane gas that is harmful to the atmosphere.

Therefore, it is my opinion that the air electric system is not only superior to hydroelectric dams but is the only viable and sustainable way of providing this earth with electrical power without creating any further gases to affect the atmosphere around us.

www.ingramcontent.com/pod-product-compliance
Lightning Source LLC
Chambersburg PA
CBHW040234220526
45473CB00001B/241